YOUR KNOWLEDGE HAS

Sebastian Wagner

Aging aircraft. Fleet planning and maintenance

GRIN Verlag

Bibliografische Information der Deutschen Nationalbibliothek:

Die Deutsche Bibliothek verzeichnet diese Publikation in der Deutschen National-
bibliografie; detaillierte bibliografische Daten sind im Internet über http://dnb.d-
nb.de/ abrufbar.

Imprint:

Copyright © 2014 GRIN Verlag GmbH
Druck und Bindung: Books on Demand GmbH, Norderstedt Germany
ISBN: 978-3-656-65189-5

This book at GRIN:

http://www.grin.com/en/e-book/272855/aging-aircraft-fleet-planning-and-mainte-
nance

GRIN - Your knowledge has value

Der GRIN Verlag publiziert seit 1998 wissenschaftliche Arbeiten von Studenten, Hochschullehrern und anderen Akademikern als eBook und gedrucktes Buch. Die Verlagswebsite www.grin.com ist die ideale Plattform zur Veröffentlichung von Hausarbeiten, Abschlussarbeiten, wissenschaftlichen Aufsätzen, Dissertationen und Fachbüchern.

Visit us on the internet:

http://www.grin.com/

http://www.facebook.com/grincom

http://www.twitter.com/grin_com

Aging Aircraft

Fleet Planning & Maintenance

Sebastian WAGNER

Maintenance Management

AVIMA 12

January 2014

Table of Contents

1 Introduction

"Public expectation is that aging airplanes are receiving increased attention to maintain safety, including airworthiness of repairs, alterations, and modifications"

(Federal Aviation Administration *(FAA) 2003, p. 3)*

Indeed, the majority of airlines are faced with the challenge of aging fleets and when it might be optimal to replace older aircraft. Well, any discussion of the wisdom of retaining capital equipment is usually based on **economic arguments**. In a competitive environment, airlines are continuously obliged to improve their business and equipment to stay profitable. The prediction of future maintenance costs of the own fleet is an integral element of prospective budgeting projections; on the other hand they serve as a vital part within aircraft replacement calculations. For example if the costs of maintaining the existing equipment on a timely basis exceeds the capital, interest, and amortization charges on replacement equipment, the decision to buy a sort of replacement is straightforward. In most cases the substitute equipment even offers an improved productivity as well (*Dixon 2006, p. 1)*.

Beside any debate concerning costs and efficiency, **flight safety considerations** also enter into the discussion especially in the field of aviation. The question to repair or replace is an ongoing decision making process for the maintenance department of every airline operator.

Now the key questions to be answered in this context are: Is it possible to describe a standard airplane service life and how does the fleet age of world's leading airlines look like? How does the process of maintenance develop over an aircraft's whole life cycle and can necessary costs be estimated? What can be done technically to keep aging effects of aircraft under control and when might be the right time to withdraw an aircraft from service?

In order to answer the abundance of questions my term paper is divided into an economic based part including compiled data and statistics and a more technical part. In the beginning, this paper investigates the ordinary economic life of commercial airplanes. Additionally I'm going to inspect exemplary the average fleet age of world's leading airlines. In the second stage I am going to describe how to estimate maintenance costs of aircraft that grow older. Further I wanted to clarify technical aspects and problems that might occur more frequently with the rising age of an aircraft.

2 Fleet Planning & Development

The planning process of an aircraft fleet for an airline is basically not different from any other planning activity in an economic environment. To establish a successful and inherent fleet plan, it requires a well mixture of commercial knowledge, engineering know-how, the ability to predict the future, and as always some luck (*Clark 2007, p. 1 f.*). Within the process of fleet planning certain knowledge of the term 'airplane economic life' is of high interest for airlines, manufactures and various other interest groups in order to predict future demand of new aircraft. In this Chapter I will go into more detail by investigating the ordinary economic life of commercial airplanes. In a second step I'm going to examine the average fleet age of world's leading airlines and will draw conclusions from it.

2.1 The Economic Life of an Aircraft

In industry literature a variety of terms can be found describing the life expectancy of an aircraft. It doesn't matter whether one favours the term 'airplane useful life', 'airplane service life', or even uses a host of variations on these terms. All explanations, however, are struggling to precisely define these terms in a way that can be quantified. There is simply no industry standard to quantify the economic life of modern commercial aircraft. Nevertheless, a commonly used metric to describe the life expectancy of an aircraft is the **average age** of an airplane when it is **permanently withdrawn from service** (*Jiang 2013, p. 2 f.*).

By the end of 2012, the aviation industry appeared to be headed towards a significant drop in the annual average age for commercial passenger jet aircraft retirements as depicted on the next page in Figure 1. A well tended database source shows an **average retirement age of 23.2 years** in 2012, which is equivalent to the lowest level in more than two decades. This trend might have potentially big implications for aircraft residual values, financing options, parts pricing, lease rates and more (*Compart 2013*). It is important to now that the economic lifespan of aircraft is always **crucial to investors** who will need to depreciate their assets and the subject is perhaps more relevant now than ever. One reason for the recent decline in aircraft retirement age might be the market entry of the so called 'Next Generation Aircraft' programmes like 737NG and A320E, enforced age restrictions, and order books burst with a record number of orders (*Baldwin 2013*). Other sources like manufacturer Boeing are blaming the great recession, a weak cargo market, and parting-out some young airplanes for the decline in average age of retired airplanes. Furthermore Boeing argues that the introduction of current-generation airplanes is in line with historical trends of their predecessors and that no impacts on the economic life of the worldwide fleets can be expected (*Jiang 2013, p. 3*).

A trend to shorter economic lives for passenger aircraft would increase the availability of good quality candidate aircraft for conversion. So called **'P2F programmes'** are otherwise common methods of prolonging the life cycle of an aircraft. Converting veteran passenger airplanes into freighters economically results in savings for cargo airline customers in comparison to new build cargos aircraft. According to the latest developments by the end of 2013, P2F conversions are set to continue as the primary source of capacity for the air cargo market (*Harris 2013*). A fairly interesting illustration of how a fleet develops as airplanes are added and removed is attaching to this paper in Appendix 1.

However, the recent **trend of dropping retirement age will not last**. The Irish aircraft lessor Avolon notes that the Airbus A320 family and 737 Classic fleets are relatively young, so airlines retiring the oldest and least efficient of those aircraft are driving down the average. By the time the airlines finally retire the younger ones, Avolon says, most will exceed 25 years of age. Additionally the company supports the thesis that the in-service lives of core single and twin aisle fleets are not experiencing rapid material deterioration. They take the view, that the industry's economic life assumptions and depreciation policies will remain valid over the next decade and beyond (*Avolon 2012, p. 1*).

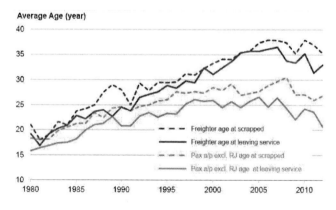

FIGURE 1: AVERAGE RETIREMENT AGE OF PASSENGER AND FREIGHTER AIRCRAFT (*JIANG 2013, P. 6*)

The Graph above depicts the trends of the average age of commercial passenger and freighter airplanes that have been permanently removed from service. The solid lines represent the average age at which airplanes leave service. The dashed lines are symbolizing the typical age at which airplanes are scrapped. The difference between solid lines and dashed lines reflects the time that airplanes are in storage prior to being scrapped. Today there are approximately 4.000 aircraft parked or stored in short- or long-term programmes (*Baldwin 2012*). What can also be seen in the graphic is that **average age at end of service has remained stable for more than 15 years**, gradually increasing as technology advances have been implemented. Over time, significant events, regulation and technology shifts have had impactful influence on the data, although often over only limited periods. These help to explain some of the variation and data.

2.2   Airline Fleets – A Market Overview

Based on the finding of the preceding Chapter, I want to take a deeper look on the worldwide fleet development and average fleet size / age of world's leading airlines.

According to an internal research of Boeing a total amount of **31.032 commercial jets were built and delivered** by western manufacturers since the start of the jet age in 1952 (*Jiang 2013, p. 3*). As if this wasn't enough, Boeing is able to predict that the commercial fleet size in service will grow with an average of 3.6 percent per year to **double its size** from 20,310 airplanes today to **41.240 within the next 20 years**. That means that the airline industry will need 35.280 new airplanes, of which 41 percent will replace older, less efficient airplanes. Furthermore an estimated amount of 6.000 to 8.000 mostly wide-body aircraft are expected to be retired worldwide in the next 10 to 20 years (*Boeing 2013a, p. 15; Tegtmeier 2007*).

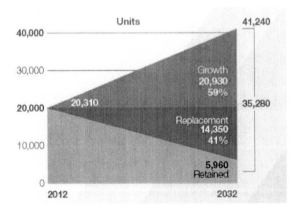

FIGURE 2: FLEET DEVELOPMENTS UNTIL 2032 (*BOEING 2013A, P. 15*)

Without a doubt, these numbers are very impressive. But where actually is this huge demand of new aircraft coming from?

In my opinion the development is based on two reasons. On one hand an increase of the **annual global air passenger traffic demand** of about 5 percent is expected until 2032. That equals a total amount which is **more than 2.6 times larger than in 2012** and carriers are forced to increase their fleet size to serve the emerging market (*Boeing 2013a, p. 32*). On the other hand the average fleet age of world's leading carriers (in regard of seat numbers offered) is fairly high in comparison to smaller aspiring airlines. In other words, to defend their market position and to reduce operating costs, the global airline players will significantly renew their fleets by introducing latest aircraft technologies within the next decades (*Clark 2007, p. 11*).

To support my theory I conducted a brief market survey based on latest data. In my own interest I considered and compared a selected choice of the largest carriers on the global aviation market in regard of fleet size and their respective aircraft age. The results are presented in the illustration below.

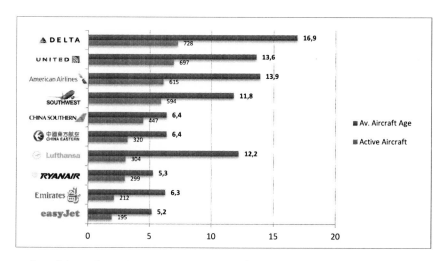

FIGURE 3: AVERAGE AIRCRAFT AGE IN COMPARISON TO ACTIVE FLEET (*OWN ILLUSTRATION 2014; AIRFLEETS AVIATION 2014*)

As of January 2014 the Delta Airlines mainline fleet comprises 728 aircraft in service. Excluding stored aircraft the average age of the Delta fleet is 16.9 years. The oldest airplanes in the fleet are the McDonnell Douglas DC-9s and the MD-88s averaging 35.1 years and 23.2 years of longevity. As we can see clearly, other U.S.-based airlines, which are operating the largest fleets worldwide, have a quite similar elderly age distribution. In comparison to operators from China and even aspiring Low Cost Carriers (LCC) from Europe their fleet structure seems really outdated. To make a long story short, world's leading airlines are facing constraints to regenerate their core fleets in the foreseeable future. In my opinion, staying competitive in the global environment means to use latest resource saving aircraft technologies. The last aviation trade fair was leading the way already. The Dubai Airshow (Nov 2013) has seen aircraft sales shoot through the roof even just after the first day. Alone the world's two biggest plane-makers, Boeing and Airbus, accounted for about $179 billion in combined orders (*Mutzabaugh 2013*). Well, the trend seems to come true as some of the American dinosaurs are going to rejuvenate their outdated fleets comprehensively. Apart from over 600 aircraft in service, American Airlines has **more than 450 aircraft** from Airbus and Boeing on order, intended to replace its aging MD-80 series, 757-200 and 767-200 jets (*AMR 2011*). The airlines United and Southwest try to be second to no one as well. Both are going to upgrades their fleets with 150 (United) and 200 (Southwest) brand new aircraft (*Boeing 2013b*).

As we can see, a great number of airlines have decided to eminently rejuvenate their fleets. Within the decision process every single aircraft operator needs to determine when it is the right time to replace an aircraft because of age. Unfortunately the problem lies in the fact that there is **no set answer** to that. Basically an aircraft has no age limit. If the proper maintenance procedures are followed, chronological age is not a limitation. For example the average fleet age in Venezuela amounts to 35.2 years (*Fessler 2012*). As we know the average retirement age of aircraft are around 23 years. Indeed, fleet ages of 35 or more are fortunately no international standard, even if there are no information about recent incidents exist. However, as long as inspections are conducted in regular intervals and components are replaced according to their calendar, the lifetime of an aircraft is only limited by the sky.

3 Maintenance of Aging Aircraft

The argument that older equipment is more expensive to operate and maintain is intuitively appealing. Within this chapter I will try to clarify whether there is a relationship between the age of equipment and the associated costs of operating and maintaining equipment. Additionally parameters will be clarified which indicate that it is time to replace an aircraft. In a second step I will now go in more detail by describing technical aspects and problems that might occur more frequently with the rising age of an aircraft.

3.1 The Effect of Age on Total Maintenance Costs

It is common sense, that proper aircraft maintenance is essential for keeping an aircraft and its parts in optimal condition and ensuring the safety of pilots, crew, and passengers. Maintenance work is a comprehensive, ongoing process and aircraft are required to be maintained after a certain period of calendar time, flight hours or flight cycles (*Kinnison/Siddiqui 2012, p. 3 ff.*).

The determination of financial resources connected to the conduction of maintenance might be **one of the most difficult disciplines** from a cost point of view. Maintenance costs are incurred during necessary procedures to retain or restore any piece of equipment or system to the specified operable condition to achieve the maximum useful life of an aircraft. That includes for example the verification of airframe systems, engine functions, replacement of worn or defective components and even the management of unscheduled failures of a system (*Clark 2007, p. 175*). Within the economic lifetime of an aircraft all these inspections are conducted in prescribed periodic intervals that have to be fulfilled after a certain amount of time or usage in compliance with the respective Airworthiness Directives or Service Bulletins.

Based on the findings of several comprehensive studies conducted by aircraft manufactures like Boeing or even independent researches, one can say that the **'maintenance life' of an aircraft can be split into three stages.** The separate stages are covering the aircraft with an average age of six years or less, aircraft older than six but no more than 12 years old, and finally aircraft older than 12 years (*Dixon 2006, p. 13 ff. & p. 25 / Clark 2007, p. 179*). Time spans defined in other surveys do not vary materially different.

The first stage (0-6 years) can be defined as **'honeymoon period'**. Within this phase the aircraft operator profits from significantly low maintenance expenditures – less than half of what they are at year five or six. The reason is a combination of the newness effect of equipment, particularly of engines, and secondly warranties of the manufactures are still in effect (*Wynham 2013*). In theory, a commercial aircraft owner should not face substantial maintenance costs in its first three to five years of ownership. For sure, these advantages need to be balanced by additional costs as mechanics for example proceed down in the learning curve in case of a brand new aircraft type in the fleet. The aircraft maintenance costs, however, will be lower than for mature aircraft of 6 or more years (*Clark 2007, p. 178*).

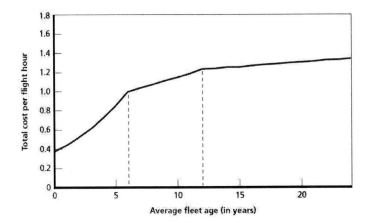

FIGURE 4: EFFECTS OF AGE ON TOTAL MAINTENANCE COSTS (*DIXON 2006, P. 28*)

As we can see in the Figure above there are three separate and independent log linear regressions dedicated to each stage of maintenance costs. The increasing aging effect in the first six to seven years results, as prescribed, mainly from warranty expiration and ends with the first D-Check* (*Dixon 2006, p. 13*).

The second stage can be described as **'mature period'**. At the starting point maintenance costs are normalized (1.0) and increasing with a moderate annual age effect. In other illustrations/surveys the average maintenance cost curves seem to level within this stage, at least for certain period (*Dixon 2006, p. 14; Clark 2007, p. 179*). Within this stage the aircraft and the associated equipment runs and can be maintained without special abnormality. Maintenance cost factors are moving around the normal cost level per flight hour. The second period usually ends with the second D-Check (*Dixon 2006, p. 13*).

---

*
A D-Check is complete structural check and restoration

During the course of life of an aircraft, the third period in regard of maintenance starts when the airplane begins to show signs of incurring higher than at mature level. This last time frame can be therefore called **'aging period'**. Put simply, maintenance costs start to rise again from around 12 to 14 years beyond entry into service. An interesting aspect is that a steadily increase in maintenance costs of a post-12-years-old aircraft can be forecasted for the third stage as shown in Figure 4. Based on several independent investigations it was found out, that with every additional year of age the costs of operating and maintaining aircraft are roughly increasing by 1 to 3 percent until the end of the aircrafts operational life (*Kiley 2001, p. 2; Dixon 2006, p. 27*). Reasons for the steady increase are perfectly obvious. As with any mechanical device, the increased maintenance (parts and labor) is primarily due to unscheduled repairs. As an aircraft ages, the increase in unscheduled maintenance associated with scheduled inspections means an increase in total maintenance down time as well. Statistical data proves that the aircraft availability can drop from the 95% range for aircraft up to 15 to 20 years of age to an average of 70% at age 25 and 55% at age 30. Looking at it the other way around, it typically takes two older aircraft to have the same availability as one newer one. Coming back to the maintenance cost based view, it is proven that an almost 30 years old aircraft is typically more than double times costly in comparison to an aircraft at the age of 5 (*Wynham 2013*).

However, when an airline decides to operate a fleet which consists of primary elderly aircraft, the company needs to adopt precautionary measures in order to keep rising maintenance costs under control. Such a strategy might include maintaining several fleets with overlapping capabilities to minimize the risk of early retirement, developing spare maintenance capacity to ensure against the effects of unexpected maintenance workload growth, and finally contingency plans for replacing older fleets (*Pyles 2003, p. 26*). From an operators view there is **unfortunately no clear formula** that announces when the lifetime of an aircraft is expired and it should be withdrawn from service. Instead, operators need to track the three key parameters mentioned in the pages above. Total maintenance costs, aircraft availability and mechanical reliability. At some point, these parameters will indicate that it is time to replace the aircraft. Nevertheless, the answer of 'when' will be different from operator to operator depending on the aircraft's mission and how much they use it (*Yang 1980, p. 528; Wynham 2013*).

Technically seen, an aircraft's lifespan is measured not in years but in pressurization cycles. Every time an airliner takes off, is pressurized during flight, and lands it goes through a demanding cycle of stress reversals. On average, an ordinary short-haul aircraft does that half a dozen times a day – year in, year out. To a varying degree, **all materials** likely to be used in aircraft structure **are subject to fatigue in one form or another**. Structural fatigue, which means a progressive loss in strength under cyclical loading, is perhaps the major issue for ageing aircraft. The age itself, however, is not directly causing the problem. It is rather a correlate of many other material-deterioration and maintenance-response processes that change over time (*Phelan 2011*). As a result, every aircraft which exceeds a certain time of life either have encountered, or can be expected to encounter, aging problems such as **fatigue cracking, stress corrosion cracking, corrosion, and wear**.

A basic problem might be that the realized life of an aircraft is **seldom equal to the design life planned for a fleet**. Because from a profit orientated operators view, the life of an aircraft fleet is generally determined by its inherent operational capability and maintenance costs and not by the number of flight hours or cycles specified at the design stage. Finally fatigue cracks will grow in the same way as the population of aircraft ages. A key problem in determining the crack status for example in an airframe or turbine blade is, that an hairline crack might remain undetected even after inspections due to the uncertainties of the inspection process (*Behrens/Burns/Rudd 1991, p. 38*). **Nondestructive evaluation (NDE) inspections** are a valuable tool in detecting those hairline cracks. They are used during an aircraft's service life to detect cracks as small as 0.04 inch and even during the production (to ensure that components start out free of defects) of an aircraft. Inspectors might take a close look at fastener holes located at the wing and spar junction for example. One commonly used method of NDE is ultrasonic phased-array testing. This method analyzes the echoes from ultrasonic waves to reveal imperfections inside a material. By using several ultrasonic beams instead of just one, inspectors can look inside a material at different locations and depths, thereby determining the size and shape of any defects (*Maksel 2008*).

Since we know that structural inspections can get quite complex and in many areas cracks and corrosion can be very difficult to find, the limitations on very high cycle airplanes that have exceeded their design life are increased. While they can still be flown safely, the inspection criteria are complex and require a lot of engineering resources to be analyzed. Regardless, in managing aging fleets, **decisions must be made concerning timing and extend of inspections, repairs, modifications and life extension options** (*Behrens/Burns/Rudd 1991, p. 37 f.*).

The resulting challenge for the technical management is to meet the following objectives (*Tiffany et al. 1997, p. 13*):

- Identify and correct problems that could threaten **aviation safety**.
- **Prevent or minimize problems** that could become an **excessive economic burden** or adversely affect airline readiness.
- For the purpose of future fleet planning, **have the methodology to predict** when the maintenance burden will become so high, or the aircraft availability so poor, that it will no longer be viable to retain the aircraft in the fleet.

Most current fatigue life assessment methodologies for advanced metal and composite structures are based on several decades of experience in full-scale structural certification and service. Nowadays the most common methods rely on empirical **stress to number of cycles (S/N) data**. A simplified illustration, so called 'Woehler Curve', is shown in Figure 5. Variations of material characteristics, lay-up configurations, loading conditions, environmental impacts, etc., often make the analysis and testing enormously challenging (*FAA 2011, p. 2*).

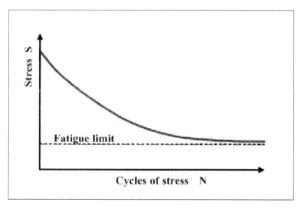

FIGURE 5: S – N CURVE – FATIGUE LIMIT (*KOPELIOVICH 2012*)

Whenever we are talking about aircraft maintenance, flight safety considerations always enter into the decision to repair or replace. Fortunately, inspection and maintenance procedures and the structural maintenance programs have been developed to reduce the likelihood of structural failure during the design service life. As a consequent development, U.S. based airlines are obliged to establish a FAA-approved structural maintenance program since the turn of the year 2010. This program is part of **'The Aging Aircraft Safety Act'** passed by the Congress of the United States in 1991. The act requires air carriers to demonstrate that maintenance of an airplane's age-sensitive parts and components has been adequate and timely enough to ensure the highest degree of safety (*Pillo/Hoggard 2010, p. 15 f.*). If we recall the FAA statement from the very beginning the wheel has come full circle now.

Other countries go one step further. Several local aviation authorities do persist on **age limitations for domestic airlines**. The list of countries establishing age restrictions on aircraft is continuing to grow. Age limitations have become much more widespread over the past four years as e.g. Russia (15-year age limit), China (10), Iran (10), Nigeria (15), India (15), Bolivia, Uganda, and many other developing countries adjust existing age limitations or impose new age caps on locally operated aircraft. Some states also draw the line at a certain percentage completed of the design economic life or even state a specific number pressurization cycles, whichever is earlier. In contrast to the aforementioned countries, there are no age limits in place for aircraft in Europe, Canada, Australia and the U.S. These regions have the most stringent aircraft regulations in place globally and already have rigorous maintenance checks in place. Whereas critical comments say that, it should safely be able to reach its design life and calendar **as long as the aircraft are operated within limits and are suitably maintained** (*Sanders 2011*).

Summarizing one can say that adequate maintenance requires the participation and cooperation of aircraft manufacturers, regulatory authorities, operators, owners, and maintainers. Conducting additional analysis on ageing aircraft and rectifying unexpected problems belongs to the responsibility of manufacturers who need to ensure continued airworthiness. Owners, operators, and maintainers need to adequately maintain their aircraft and have to report defects to the manufacturer and regulatory authorities, who in turn, must ensure that safety-related maintenance information is disseminated quickly among other operators of the type.

4 Conclusion

The aim of this term paper was to answer the key questions asked in the beginning. I asked whether a standard economic life of a commercial passenger aircraft can be described and how does the fleet age of world's leading airlines look like. During my research I was able to reveal, that the average service life of commercial aircraft is of high interest not only for manufactures and customers. It also has potentially big implications for aircraft residual values, financing options, parts pricing, leasing companies and so on.

Additionally a short survey, based on the latest fleet information, has shown that world's leading airlines are facing constraints to regenerate their core fleets to meet the increasing requirements of global air passenger traffic. To use the latest resource saving aircraft technologies will be the key to success in such a highly competitive environment.

In the second part of this assignment I tried to find out, whether there is a relationship between the age of equipment and the associated costs of operating and maintaining equipment. This paper found that aging has a profound impact on maintenance costs. On top of that, costs do change over time as they rise substantially once aircraft come off the manufactures warranty. Fortunately recent research allows predicting future maintenance costs and helps to understand and calculate the 'true costs of ownership'.

In a last step it was shown impressively that structural inspections to prevent fatigue / corrosion cracking, corrosion, and wear can get quite complex and that they are the most challenging tasks in maintaining aged aircraft companies have to deal with. However, it was shown to some extend that the lifespan of an aircraft can be determined by simple age or by more meaningful flight cycles and the number of flight hours. The respective aging limits are established and supervised by manufacturers and the respective aviation authorities. In my opinion there is nothing to worry about if you take a seat in a more mature aircraft. As long as you can be sure that all required maintenance procedures are followed, chronological age is not a limitation. Current and future maintenance programmes can act as a preventative measure to reduce the safety risk associated with ageing aircraft. In my opinion even without age restrictions, noise restrictions tend to have the effect of culling the older planes out of fleets, particularly in Europe. Finally, the ultimate product purchasing decisions will also highly depend on customer preferences and acceptance.

5 Appendix 1

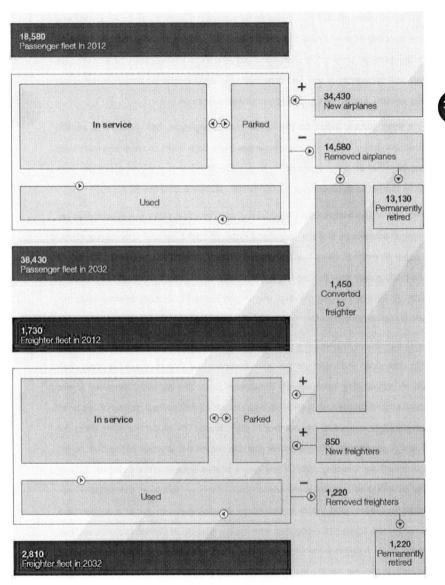

6 Appendix 2

6.1 List of Abbreviations

Abbreviation	Explanation
FAA	Federal Aviation Administration
LCC	Low Cost Carrier
NDE	Non-Destructive Evaluation Inspections

TABLE 1: LIST OF ABBREVIATION

6.2 Bibliography

Airfleets Aviation (2014): *Airline Fleet Age*, downloaded from: http://www.airfleets.net/ageflotte/fleet-age.htm, assessed on 03.01.2014

AMR (2011): *AMR Corporation Announces Largest Aircraft Order in History with Boeing and Airbus*, downloaded from: http://hub.aa.com/en/nr/media-kit/fleet/landmark-fleet-transaction, assessed on 15.01.2014

Avolon (2012): *Avolon issues analysis of economic life of commercial jet aircraft*, downloaded from: http://www.avolon.aero/pressrelease/19_Sep_2012.pdf, assessed on 06.01.2014

Baldwin, M.-A. (2012): *Aircraft storage: The costs and considerations*, downloaded from: http://www.afm.aero/magazine/trading-legal-and-finance/item/285-aircraft-storage-the-costs-and-considerations, assessed on 06.01.2014

Baldwin, M.-A. (2013): *Aircraft economic life: Is age just a number?*, downloaded from: http://www.afm.aero/magazine/trading-legal-and-finance/item/570-aircraft-economic-life-is-age-just-a-number, assessed on 06.01.2014

Berens, A. P. / Burns, J. G. / Rudd, J. L. (1991): *Risk Analysis for Aging Aircraft Fleets*, published in: Atluri, S. N. / Sampath, S. G. / Tong, P.: Structural Integrity of Aging Airplanes, 1st Edition, pp. 37-51

Boeing (2013a): *Current Market Outlook 2013 –2032*, downloaded from: http://www.boeing.com/assets/pdf/commercial/cmo/pdf/Boeing_Current_Market_Outlook_2013.pdf, assessed on 06.01.14

Boeing (2013b): *United Airlines announces historic order of 100 737 MAXs and 50 Next-Generation 737s*, downloaded from: http://www.newairplane.com/737max/customers/united-airlines/, assessed 15.01.2014

Clark, P. (2007): *Buying the Big Jets: Fleet Planning for Airlines*, 2nd Edition, Ashgate

Compart, A. (2013): *2013 Could Be Telling For Aircraft Retirement Trend*, downloaded from: http://www.aviationweek.com/Article.aspx?id=/article-xml/AW_02_11_2013_p18-544654.xml&p=1 assessed on 06.01.2014

Dixon, M. (2006): *The Maintenance Costs of Aging Aircraft, Insights from Commercial Aviation*, 1st Edition, Rand Corporation

FAA (2003): *Supplemental Structural Inspection Document (SSID) Standardization Public Meeting*, downloaded from: http://www.faa.gov/aircraft/air_cert/design_approvals/transport/aging_aircraft/media/faapitch.pdf, assessed on 05.01.2014

FAA (2011): *Determining the Fatigue Life of Composite Aircraft Structures Using Life and Load-Enhancement Factors*, downloaded from: http://www.tc.faa.gov/its/worldpac/techrpt/ar10-6.pdf, assessed 14.01.2014

Fessler, P. (2012): *Die ältesten Flotten der Welt*, downloaded from: http://www.aerotelegraph.com/die-laender-mit-den-aeltesten-flotten-der-welt, assessed on 10.01.2014

Harris, D. (2013): *Demand rises for freighter conversions*, downloaded from: http://www.airfinancejournal.com/Article/3282905/Analysis-Demand-rises-for-freighter-conversions.html, assessed 15.01.2014

Jiang, H. (2013): *Key Findings on Airplane Economic Life*, downloaded from: http://www.boeing.com/assets/pdf/commercial/aircraft_economic_life_whitepaper.pdf, assessed on 05.01.2014

Kiley, G. T. (2001): *The Effects of Aging on the Costs of Maintaining Military Equipment*, 1st Edition, Washington D.C

Kinnison , H. /Siddiqui, T. (2012): *Aviation Maintenance Management*, 2nd Edition, McGraw Hill

Kopeliovich, D. (2012): *Fatigue*, downloaded from: http://www.substech.com/dokuwiki/doku.php?id=fatigue, assessed on 13.01.2014

Maksel, R. (2008): *What determines an airplane's lifespan? - Some keep flying for decades, while others end up on the scrap heap*, downloaded from: http://www.airspacemag.com/need-to-know/NEED-lifecycles.html, assessed 16.01.2014

Mutzabaugh, B. (2013): *Boeing called 'winner' at 2013 Dubai Airshow*, downloaded from: http://www.usatoday.com/story/todayinthesky/2013/11/20/boeing-called-winner-at-2013-dubai-airshow/3650787/, assessed on 08.01.2013

Phelan, P. (2011): *Ageing aircraft – a real problem, or a diversion?*, downloaded from: http://www.aviationadvertiser.com.au/news/2011/04/ageing-aircraft-%E2%80%93-a-real-problem-or-a-diversion/, assessed 12.01.2014

Pillo, R. M. / Hoggard, A. (2010): *Complying with the aging aircraft safety rule*, published in: Aero Quarterly, No. 2, Year 2010, pp. 14-19

Pyles, R. A. (2003): *Aging Aircraft USAF, Workload and Material Consumption Life Cycle Patterns*, 1st Edition, Rand Corporation

Sanders, F. (2011): *Aircraft age restrictions jeopardise aviation financing*, downloaded from: http://www.ascendworldwide.com/2011/10/aircraft-age-restrictions-jeopardise-aviation-financing.html, assessed 16.01.2014

Tegtmeier, L. A. (2007): *We Recycle, Used airplane parts can appear in the strangest places*, downloaded from: http://www.airspacemag.com/flight-today/we_recycle.html?c=y&page=2 assessed on 04.01.2014

Tiffany, C. F. et.al (1997): *Aging of U.S. Air Force Aircraft: Final Report*, 1st Edition, Washington, The National Academies Press

Wyndham, D. (2013): *When to replace an aircraft*, downloaded from: https://www.conklindd.com/Page.aspx?hid=1127, assessed on 12.01.2014

Yang, J. N. (1980): *Statistical Estimation of Economic Life for Aircraft Structures*, published in: Journal of Aircraft, Vol. 17, No. 7, pp. 528-535

18

TABLE 2: BIBLIOGRAPHY

6.3 Figures

19

6.4 Tables

Lightning Source UK Ltd.
Milton Keynes UK
UKOW02f0322180616

276508UK00003B/66/P